Writing High-Quality Standard Operating Procedures

Writing High-Quality Standard Operating Procedures

A Practical Guide to Clear, Concise, and Correct SOPs

Atul Mathur

Copyright © 2017 Atul Mathur
All rights reserved.

ISBN: 1543279317
ISBN 13: 9781543279313
Library of Congress Control Number: 2017905292
CreateSpace Independent Publishing Platform
North Charleston, South Carolina

To my mother, the late Mrs. Bibi Rani Mathur, who herself enjoyed the art of writing and encouraged me to follow the path of "work is worship"

To Anmol, my elder daughter, who silently teaches me the concept of "purpose"

Contents

Introduction ix

Chapter 1 Importance of SOPs 1
Chapter 2 Qualities of High-Quality SOPs 7
Chapter 3 Structure and Parts of an SOP 13
Chapter 4 Right Process for Right SOPs 29
Chapter 5 Writing the Right Way 41
Chapter 6 Avoiding Common Errors 67
Chapter 7 Checking Quality with a Checklist 85

Acknowledgements 93
About Atul Mathur 95

Introduction

*Real knowledge is to know the extent
of one's ignorance.*

—*Confucius*

To make the contents of this book worthy of your time and attention, let's start by posing three key questions:

- What's the specific problem this book promises to solve?
- Is that problem really important?
- How does this book provide a possible solution?

1. What's the problem?

As you know, all organizations dealing with human health, such as pharmaceutical, biotechnology, and life-sciences companies, operate based on standard operating procedures (SOPs) to ensure the quality and safety of their products and services. SOPs also ensure compliance with applicable standards, guidelines, and regulations. Almost every activity—from cleaning a production machine to running a critical process to even saving and archiving data—is carried out based on written SOPs. And as a result, these organizations end up producing hundreds, if not thousands, of SOPs that employees at all levels have to read, understand, and follow.

Containing rich technical information, these documents are often wordy and confusing instead of concise and precise. Typically, the language used is complicated instead of simple and easy to understand. Sometimes, the contents of SOPs don't reflect the actual processes followed on the production floor, leading to quality and compliance problems.

In essence, the problem is paradoxical: Pharmaceutical, biotechnology, and life-sciences companies depend heavily on SOPs to pursue excellence in the quality of their products and services. But the SOPs themselves are not of the highest quality. As mentioned earlier, these documents are wordy, complicated, and imprecise.

But then, the next question is this: Why does it matter if SOPs are not well written?

2. Is the problem important?

Not many technical documents are as consequential as SOPs, for the contents of SOPs directly affect the way people act and behave in an organization.

If SOPs are not concise, easy to understand, and precise, users may not fully understand and follow the procedures, thereby creating quality and safety risks—and worse, deviations. During regulatory audits (e.g., by the US Food and Drug Administration),

SOPs invariably come under the scanner of auditors—and many of the deviations identified during audits are often related to the gaps between what is written in SOPs (or even the absence of the relevant SOP itself) and the actual actions of the users of those SOPs. And hundreds of man-hours are routinely spent defending ineffective SOPs during audits and implementing corrective and preventive actions stemming from SOP-related issues.

By connecting the dots between SOPs and a company's operations, it is not too hard to see that just like the quality of people, equipment, and materials, the quality of SOPs does matter. If these documents are not well written, the price paid is adverse impact on organizational productivity, quality, reliability, and compliance.

If the quality of SOPs is really important to your organization, what can you do to improve it?

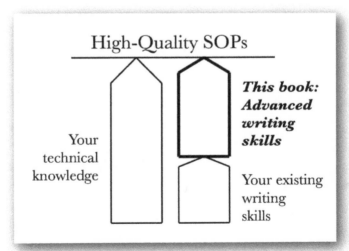

Figure 1: Advanced Writing Skills for High-Quality SOPs

3. What's the solution?

Professionals working in pharmaceutical, biotechnology, and life-sciences companies who write SOPs are technically sound and know their stuff well, but they are often not well trained in technical writing skills. The proficiency of their writing skills usually falls short as compared to the depth of their technical knowledge. Basically, they are subject-matter experts but not experts in technical writing.

As a result, despite their sound technical knowledge, they end up writing SOPs that are not as concise, clear, and precise as they can and should be.

The objective of this book is to enhance the writing skills of the professionals who write SOPs by providing them advanced writing tools and tips. And this will, in turn, hopefully help them uplift the quality of SOPs in their organization.

4. Roadmap to high-quality SOPs

CONTAINING SEVEN CHAPTERS, this book provides a step-by-step roadmap for writing high-quality SOPs.

We start the journey with chapter 1 by refreshing our understanding of the importance of SOPs. Chapter 1 specifically answers this question: Why do SOPs matter? If you understand well the impact SOPs have on product quality, compliance, reliability, and productivity, chances are you would be more willing to improve your writing skills.

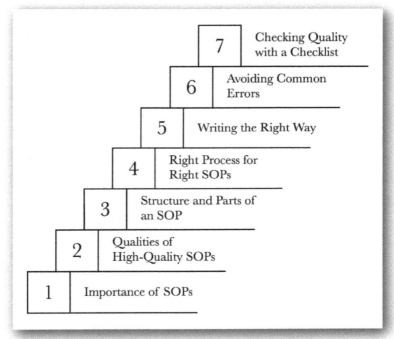

Figure 2: Step-by-Step Roadmap with Seven Chapters

Chapter 2 defines the attributes of well-written, high-quality SOPs. This chapter sets the standards to aim for while writing SOPs.

Chapter 3 guides you through the structure of SOPs. Since SOPs include substantial technical information, these documents need to be well structured if readers are to navigate them smoothly.

Chapter 4 focuses on the process of writing SOPs from beginning to the end. As the quality of any product largely depends on the quality of the process followed to produce it, following the right process for writing SOPs is important.

Chapter 5 introduces you to the best practices in writing that you can apply to enhance the quality of your writing and SOPs.

Chapter 6 highlights some of the common errors in the English language that should not be allowed to creep into your documents.

Chapter 7 provides you a simple, but powerful tool—a checklist—for systematically reviewing and checking the quality of your SOPs.

Let the journey toward high-quality SOPs begin.

CHAPTER 1

Importance of SOPs

The important thing is not to stop questioning.

—Albert Einstein

1. Objective

Professionals working in pharmaceutical, biotechnology, and life-sciences companies often have to deal with so many SOPs within a limited time that they end up writing and reading these documents only to satisfy the necessary formalities. And that's where lies the risk of forgetting the very purpose of SOPs.

What may look like a redundant question is an important starting point to begin our efforts to write high-quality SOPs: Why do SOPs matter?

The objective of this chapter is to shine light on and emphasize the benefits of SOPs. It is a gentle reminder to give SOPs their due importance.

Thought exercise (3–5 minutes)

Imagine a pharmaceutical manufacturing company that doesn't have SOPs. It has machines, materials, people, products, and clients. The employees in this company know what to do based on hands-on training, but no written procedures are available. Unthinkable? Yes, but just imagine it.

What do you think will be the problems faced by the clients, employees, and vendors of such a company?

2. Benefits of SOPs

If you did the exercise above, you will be able to better appreciate that SOPs provide the following benefits:

- **Consistency:** SOPs eliminate variations in procedures and processes followed by different employees in different plants and locations, thus promoting consistency in critical activities.
- **Best practices:** Through SOPs, organizations can implement proven and correct procedures, perpetuating "best practices."
- **Training:** SOPs provide a sound foundation for the training of users.
- **Continuous improvement:** SOPs enable control of processes, which in turn provides opportunities for continuous improvements.
- **Regulatory compliance:** SOPs minimize the risks of regulatory violations, product liabilities, and recalls.

- **Quality:** With all of the above benefits, the final outcome is superior quality of processes followed, leading to superior quality of products and services.

Overall, SOPs provide clear written guidance on how to perform important activities with consistency, which in turn helps to improve quality and compliance and to reduce risks.

Figure 3: Benefits of SOPs

3. SOPS IN THE CONTEXT OF REGULATIONS

What is the significance of SOPs in the context of regulatory bodies like the US Food and Drug Administration (FDA) that set the standards for the pharmaceutical, biotechnology, and life-sciences industries? Following is a short primer on some of the regulations that touch upon the requirements of SOPs:

Although the word "SOP" doesn't appear as such, FDA's regulations imply the need for SOPs by asking for "written procedures that accurately describe and detail essential job tasks."

The FDA's 21 *CFR* 211.100 (Written procedures, deviations) states:

> There shall be written procedures for production and process control designed to assure that the drug products have the identity, strength, quality, and purity they purport or are represented to possess. Such procedures shall include all requirements in this subpart. These written procedures, including any changes, shall be drafted, reviewed, and approved by the appropriate organizational units and reviewed and approved by the quality control unit.
>
> Written production and process control procedures shall be followed in the execution of the various production and process control functions and shall be documented at the time of performance. Any deviation from the written procedures shall be recorded and justified.

Medical device SOPs are regulated under 21 *CFR* 820, which governs the methods used in, and the facilities and controls used for, the design, manufacture, packaging, labeling, storage, installation, and servicing of all finished devices for human use. Medical-device companies must have documented procedures on each of these topics.

Beyond the written procedure, SOP compliance includes a requirement to train employees on essential job tasks, mentioned in 21 *CFR* 211.25 as follows:

> Each person engaged in the manufacture, processing, packing, or holding of a drug product shall have education, training, experience, or any combination thereof, to enable that person to perform the assigned functions.

The International Conference on Harmonization of Technical Requirements for Registration of Pharmaceuticals for Human Use (ICH) defines an SOP as "detailed written instructions to achieve uniformity of the performance of a specific function."

Overall, regulations specify a combination of written procedures and employee training to ensure the quality of drug products or medical devices being tested or manufactured.

What kind of procedures are the best candidates for standardization in the form of SOPs? Typically, processes that are repetitive, performed by different people, impact product quality, and fall within regulations (international, local, or company) are the right candidates for SOPs.

Generally, pharmaceutical companies need to have SOPs in the following areas:

- Buildings and facilities
- Equipment
- Raw-materials receipt and handling
- Production
- Drug product containers and closures
- Packaging, labeling, holding, and distribution
- Laboratories
- Records and reports
- Salvaging drug products

CHAPTER 2

Qualities of High-Quality SOPs

Quality is pride of workmanship.

—W. Edwards Deming

1. Objective

What are the key parameters that allow you to assess the quality of SOPs and differentiate between a well-written and a substandard SOP?

This chapter aims to help you identify the key attributes of well-written, high-quality SOPs.

Thought exercise (3–5 minutes)

Imagine you're visiting a pharmaceutical plant as an FDA inspector, and your job is to figure out the quality of the plant's SOPs. But there is one condition: you can't see the SOPs. What kind of observations and symptoms will tell you that the quality of SOPs in this plant is below the required standard?

Thought exercise (3–5 minutes)
Imagine you're a technician whose job is to operate, based on an approved SOP, a blending machine in a tablet-manufacturing plant. If the SOP is not written well, what kind of problems might you face during the operation?

2. Qualities of high-quality SOPs

Have you ever come across a written procedure that was too complicated to follow and didn't help you to achieve your objectives? In frustration, you likely simply gave up trying to follow it. If not at work, you might have had such an experience while trying to set up a DVD player or an electronic watch or a new program. Besides failing to help the users and instead frustrating them, defective procedures can also lead to legal, financial, and other liabilities.

In contrast, a high-quality procedure is, first of all, not overwhelming in its appearance unlike documents that present a fog of densely packed text of small fonts. Next, it is clear, complete, and precise. Furthermore, it is well organized and easy to navigate, read, and understand. In the end, it is written to help you achieve your core objective, which is to execute a procedure safely and efficiently.

Overall, the qualities of a high-quality procedure can be summarized as follows:

- **Well-formatted:** Does the appearance of a document really matter? The truth is that both the looks and substance of a document are important as the two elements complement each other. If an SOP is well formatted and neatly packaged, readers can efficiently read it and grasp the substance.

Typically, SOPs are cluttered and cramped, with small fonts and dense text, and the format (with narrow margins) is often far from reader friendly. Forget about the contents first; at the basic level, an SOP should be neat and reader friendly in appearance, so people feel like reading it.

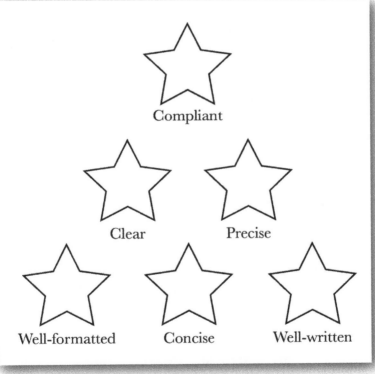

Figure 4: Qualities of High-Quality SOPs

- **Concise:** How much information should be included in an SOP?

    Many times, out of the urge to make SOPs as comprehensive (and impressive) as possible, authors of SOPs tend to include too much information, regardless of its direct relevance. But under time and work pressure, readers find such SOPs overwhelming, complicated, and difficult to follow. And that's a direct risk for noncompliance and audit issues.

    SOPs should be concise, with due respect for users' time constraints and specific process requirements.
- **Well-written:** Long sentences versus short sentences? Long and complex words versus short and simple words? Are abbreviations and jargon used but not explained? The quality of language you use directly affects the quality of an SOP itself.
- **Clear:** In the end, are readers left confused or clear about what to do?
- **Precise:** Are the various figures, units, and process steps accurate and precise?
- **Compliant:** Is the SOP compliant with the applicable internal and external quality standards, guidelines, and regulations?

Exercise (3–5 minutes)

On a scale of zero (bad) to ten (excellent), how would you rate the quality of SOPs currently used in your organization?

Well-formatted: _____
Concise: _____
Well-written: _____
Clear: _____
Precise: _____
Compliant: _____

CHAPTER 3

Structure and Parts of an SOP

The whole is more than the sum of its parts.

—*Aristotle*

1. Objective

A well-structured document makes it easy for readers to navigate and understand it without facing any obstacles. The objective of this chapter is to help you understand the overall structure of an SOP and the significance of its key parts.

2. Three-part Structure: Front, Body, End

The FDA and other regulatory bodies don't impose any standard format for SOPs. Although procedures may differ in terms of purpose, subject, and audience, in general, an SOP is a three-part document, comprised of the following elements:

- Part 1: Front matter
- Part 2: Body (procedural steps)
- Part 3: End matter

Here is what goes into each of the three sections.

Front matter

- Title page (including SOP number, author, reviewer, approver, dates, revision history, logo, and so on)
- Table of contents
- Purpose
- Scope
- Responsibilities
- Definitions
- Abbreviations
- Materials, equipment, machine
- Cautions, warnings, and dangers

Body (action steps)

This is the heart of an SOP where you include the step-by-step instructions.

- First step
- Second step
- Third step and so on

Sometimes, each step may also include the following subpoints:

- Purpose
- Materials, equipment, and special conditions for this step
- Substeps

END MATTER

- Contingencies, corrective actions
- References to other SOPs
- References to internal and external regulations
- Attachments (checklist, forms, and so on)

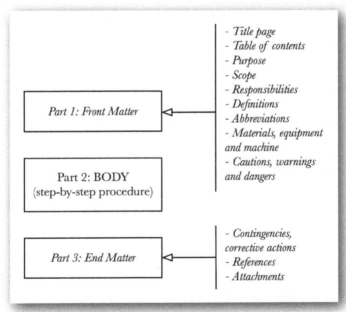

Figure 5: Three-Part Structure of SOPs

3. Key parts

Let's look at each of the parts in more detail.

Title page

The title page includes the following information:

- Name of the organization
- Short title of the procedure
- SOP number
- Date of issue
- Revision history
- Signatures (with dates) of those individuals who write, review, and approve the procedure

Table of contents

Include a table of contents if it is a long procedure.

Purpose

What the procedure is for? Clearly state here what readers can accomplish by following the SOP. Here are a few examples:

- *The purpose of this standard operating procedure is to provide a plan of action for releasing a test sample to a US Food and Drug Administration (FDA) inspector during a plant investigation.*

- *This procedure defines the method for conducting filter-integrity tests in production clean rooms.*
- *This SOP explains the procedure for transporting an islet product to the clinical-infusion site for a recipient patient.*

Keep this section short and sharp; a few sentences are sufficient.

Scope

Every SOP should define its boundaries in terms of activities, people, departments, equipment, or locations. If scope is not clearly defined, users may apply an SOP in inappropriate situations. Here are some questions to ask while defining the scope:

- Which specific departments does this procedure apply to?
- Which specific groups of people does this procedure apply to?
- Which specific process and machine does this procedure apply to?
- Which specific materials or products does this procedure apply to?
- Are there any specific capacities or volumes to which this procedure applies?
- What does it exclude?
- Are there any limitations or exceptions?

Responsibilities

Under this section, identify the personnel, departments, and contractors responsible for complying with an SOP. There can be several people (e.g., production coordinator, production operator, supervisor) and departments (e.g., quality assurance and training departments) responsible for compliance with an SOP.

Avoid using names of specific employees and instead use job titles or functions (e.g., production planner).

The "responsibilities" section is particularly significant because during audits, inspectors are more likely to look at it and verify if the "responsible people" have been adequately trained. Ideally, companies should take care to ensure alignment between employees' training and this particular section of SOPs.

Finally, the SOP writer should consult people and departments before including them in the "responsibilities" section. This helps to avoid unnecessary resistance and improve the acceptance and implementation of an SOP.

Definitions and abbreviations

Define here all special terms and abbreviations that readers are unlikely to be familiar with. If a particular term to be included in this section is also defined in the applicable standards or regulations, use the same definition in the SOP to ensure consistency.

Materials, equipment, and machines

If a procedure includes dealing with any special materials, equipment, or machines, include them here.

Cautions, warnings, and dangers

The reader should be able to clearly understand the risks involved—that is, what might happen if the procedure is not followed—and also the precautions for safe operation. Depending on the severity of the risks involved, you can include cautions, warnings, or dangers as described below.

Caution

A caution prevents a possible mistake that could result in damage or injury. Here's an example:

Do not use any hard object against the HEPA filter's face area. This can damage the filter.

Warning

A warning alerts users to potential hazards to life or limb. Here's an example:

Fire or explosion hazards exist under certain conditions with R-134a refrigerant. A combustible mixture can form when air pressure is above atmospheric pressure, and a mixture of air and R-134a exists. For this reason, do not pressure-test air-conditioning systems with compressed air.

All containers with refrigerant are under pressure (to contain the refrigerant). Any heat will increase that pressure. The containers are not designed to withstand excessive heat even when empty and should never be exposed to high heat or flame because they can explode.

Danger

A danger notice conveys immediate danger to life or limb. Here's an example:

The red container contains a dangerous substance that could choke any nearby personnel if released into the room. Keep the lid closed at all times.

Where possible, use symbols like those used on high-voltage electricity panels to direct readers' attention to the caution or warning or danger notice.

Procedure

This is the core of an SOP that people will read and act upon to achieve their objectives, such as shutting down a machine, operating a bio-safety cabinet, or blending ingredients of a drug. This section should describe specific steps, showing how to execute the procedure in the same order as someone would actually perform it.

As supplementary information, include the following:

- Instructions for completing any supporting documents, such as forms, checklists, and so on
- If it is an elaborate procedure with multiple steps, sometimes it makes sense to include in this section items that are usually mentioned separately in the procedure (e.g., purpose or principle behind each step, applicable materials, tools, equipment, and so on).
- Necessary notes, warnings, cautions, and dangers as applicable

To write this section, think and imagine how a procedure is performed from the beginning to the end. Better still, observe someone performing it, or perform it yourself.

Avoid directly copying relevant clauses from the related regulations word for word in the SOP. A procedure should essentially remain a document for describing specific actions—and not turned into a technical specification.

Notes within a procedure

A note clarifies a point, emphasizes key information, or describes alternatives or options. The following example shows how a note supports a procedural step:

Procedural step: *Apply a preliminary test pressure of 25 psi for a minimum of ten minutes.*
Note: *If leaks are detected during this step, or at any time during the test, relieve the pressure and consult the project engineer for instruction.*

CONTINGENCIES AND CORRECTIVE ACTIONS

Identify and state here any contingencies that may arise and what the operator should do. What specific actions are needed? Whom should be informed? Which specific documents should be filled?

What kind of contingency situations we are talking about? Examples might include spilling the ingredients of a drug on the floor, a room pressure or humidity alarm going off, or foreign particles being detected in raw materials.

Likewise, state what happens when an SOP is incorrectly followed. Include both short-term and long-term corrective measures.

REFERENCES

List related SOPs, supporting documents, and applicable regulations and regulatory guidelines. Make sure all the references are valid as this is often a weak link in SOPs that gets exposed during audits.

Appendices

List applicable forms that are required to be completed as per the SOP. Attach any documents used in support of the SOP (e.g., flowcharts, work instructions, pictures or diagrams, forms, and labels). Again, make sure all attached documents are valid.

Exercise (3–5 minutes)

List three ways in which you can improve the structure of SOPs in your organization.

SAMPLE TEMPLATE

STANDARD OPERATING PROCEDURE (SOP)

Written By

Name & Designation	Signature & Date

Reviewed and Approved by

Name & Designation	Signature & Date

Revision History

Revision No. & Release Date	Brief Description of Changes

Purpose

What this procedure is for? Clearly state the objective of the procedure.

Scope

What does it include? Defines the boundaries of this procedure in terms of processes, departments, locations, equipment, and so on. What does it exclude? Any limitations or exceptions?

Responsibilities

Who is responsible for complying with this procedure? Identify here the personnel (whether internal or external or both) responsible for executing this procedure.

Definitions and Abbreviations

Include all relevant definitions and abbreviations for ready reference.

Materials, Equipment, and Machines

Which specific machines, equipment, or materials are involved? Include them here.

Cautions, Warnings, and Dangers

Should users be aware of any cautions, warnings, and dangers for their own safety? Include them here.

- Caution: A caution prevents a possible mistake that could result in damage or injury.
- Warning: A warning alerts users against potential hazards to life or limb.
- Danger: A danger conveys immediate danger to life or limb.

Procedure

What should users do to achieve the purpose mentioned in the beginning? Include step-by-step instructions. Choose the most appropriate format from the following four options:

- Simple steps: Applicable for relatively short procedures, a simple-steps format lays out a procedure in a linear way.
- Hierarchical steps: When a greater number of steps is involved, use this format, which allows you to create a hierarchy—with main steps and substeps within each main step.
- Graphical format: This format includes images with explanatory text.
- Flowcharts: A flowchart is suitable for procedures involving a number of logical steps and decisions.

Contingencies and Corrective Actions

What should users do if any contingencies (e.g., power disruption, alarms, and so on) arise? What should be the corrective actions?

References

Include a list of referenced SOPs, regulations, standards, and guidance documents.

List of Attachments

Include a list of forms, diagrams, pictures, labels, and so on that are attached to the SOP.

CHAPTER 4

Right Process for Right SOPs

I am going to lay this brick as perfectly as a brick can be laid. You do that every single day. And soon you have a wall.

—Will Smith

1. Objective

The very fact that SOPs are mandatory for all critical activities underscores a key principle of creating anything: a product and its process of production are not separate, and getting the process right is the only way to the right product. And this is true for SOPs themselves. If you focus on and follow a systematic process of writing SOPs and avoid detours and shortcuts, the outcome will be a high-quality product.

The objective of this chapter is to walk you through a step-by-step process of writing effective SOPs.

Exercise (3–5 minutes)
What is the process you follow in your company? Create a simple flowchart.

2. Who should write SOPs?

Writing SOPs may seem like a typical "documentation" work and some companies, under pressure to generate SOPs, delegate this responsibility to a few dedicated "writers" who may not fully understand the procedures they are writing about. Be aware this kind of document-churning operation is a red flag for the FDA, and such cases have attracted warning letters in the past.

SOPs should be written by personnel who have a good understanding of the procedures and who have the ability to describe those procedures correctly. These people could be actual users or subject-matter experts (SMEs). Sometimes, it may not be possible to give the whole responsibility of developing SOPs to actual users; in such cases, involving them as a part of the SOP team is a good idea. When people from different departments play active roles in developing SOPs, their buy-in to these procedures is easier and faster.

3. Process of writing an SOP

It's hard to believe, but the very process—leaving alone the writing skills—used for developing SOPs affects their quality and effectiveness. Indeed, that's the reason why many organizations actually have an SOP called "Writing SOPs," but usually, this document is limited to providing a standard format—and offers nothing specific in terms of writing tips.

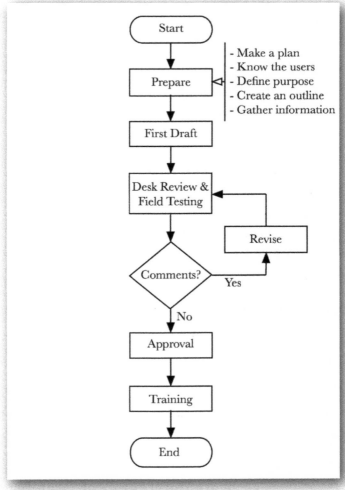

Figure 6: Process of Writing an SOP

Following a systematic process ensures SOPs are conceptualized, written, reviewed, tested, and authorized in the right sequence by the right people.

Let's look at the step-by-step process, which is comprised of the following steps:

- Prepare
  - Start with a plan
  - Know the users
  - Define the purpose
  - Create an outline
  - Gather relevant information
- Write the first draft
- Conduct desk review
- Test the prototype SOP with real action
- Review and approve
- Train the users

Next, let's see the relevance of each of the steps above.

3.1 Prepare

Paradoxically, writing an SOP doesn't start with writing; it starts with taking the following preparatory steps. Any shortcuts at this stage can hurt the quality and effectiveness of a procedure.

3.1.1 Start with a plan

Writing an SOP is a project with multiple steps that need to be executed in a specific sequence. If you start without a plan, you may underestimate the time required for developing an SOP, and then, under pressure, you may bypass or compress important steps, leading to a compromise of quality and effectiveness. So always start with a plan, which should specifically include ample time for reviewing and revising, as it is not uncommon for an SOP to go through six to ten revisions before its final approval.

Multiple reviews and revisions are the keys to the final quality and effectiveness of an SOP.

3.1.2 Know the users and empathize

One of the classic pitfalls you must avoid is ignoring the level of users' knowledge and capabilities. Before writing, step back and spend some time answering the following questions:

- Who are the target users of this procedure?
- What are their qualifications?
- What's their level of knowledge and skills in the process to be described in the SOP?
- How experienced they are?
- What may be their likely concerns?
- What are their objectives? What do they want to accomplish?
- Are there different categories of users—highly experienced and inexperienced—involved?

Ultimately, can you step into the users' shoes and empathize with them?

3.1.3 Define the purpose

Define the purpose of a procedure. What's the objective? What should the users be able to achieve?

3.1.4 Create an outline

Here's an easy way to get started: observe someone performing the procedure and note down each step taken by this person. Then create an outline or flowchart of the procedure based on your observations. If a physical demonstration is not possible, you can discuss the key steps of a procedure with a potential user and still create an outline to start with.

3.1.5 Gather relevant information

Finally, gather all the relevant information (data, specifications, regulatory guidelines, charts, photos, etc.) from colleagues, related procedures, standards, and regulations. The idea here is to simply create a resource bank before you launch into writing.

3.2 Write the first draft

Postpreparation, it's time to write the first draft. Depending on the complexity of the procedure, choose one of the following structures (more about these structures in the next chapter) to organize the steps:

- Simple steps
- Hierarchical steps
- Graphical format
- Flow chart

Don't try to be perfect with the first draft or else you may end up spending a disproportionate amount of time on this stage. The quality and effectiveness of an SOP builds during the revision phase, for which you should allow ample time. As mentioned earlier, it's not unusual for an SOP to go through six to ten revisions before its final acceptance.

3.3 Conduct desk review

Once the first draft is ready and is in reasonable shape, rope in the actual users (and also other reviewers), and ask them to review and comment. But at this stage, be cautious of the classic pitfall, which is all too common: a quick but not well-rounded response from the reviewers.

If you simply send the draft SOP to reviewers without specifying the key aspects that you want them to scrutinize, they may give two comments in the first round, three more when they look at it the second time, and a few more when they see it again, triggering an endless cycle of revisions. And still some vital aspects may be missed.

To bring seriousness and precision to the process of review, request that all reviewers necessarily comment on the following aspects:

- Is the format neat and reader friendly?
- Is it easy to read and understand?
- Are the contents well organized and logically sequenced?

- Is it concise?
- Is it clear?
- Is it precise and valid?
- Is the procedure complete in all respects?
- Anything to be added or deleted?

When people who have a stake in your SOP are involved in its development in a committed manner, they will be more inclined to own and use it.

3.4 Test the prototype SOP with real action

Even after a thorough desk review, the question still remains: Can actual users understand and follow it without any confusion and consistently produce the intended results?

One proven way to ensure the quality and effectiveness of an SOP is to test it out in the field—before it is approved. Let the users try out the procedure by performing each step exactly as described while the procedure writer observes. If the person carrying out the trial is someone unfamiliar with the procedure, you may get even better feedback on the effectiveness of an SOP. (Fresh eyes can often observe things that trained ones can't.) Lastly, it also helps if the people writing SOPs sometimes go into the field and try the SOPs out themselves. During the trial, specifically look for these elements:

- Missing steps
- Illogical steps
- Unnecessary steps

- Hesitations and confusion points
- Risks and dangers
- Accuracy of results
- Consistency of results

For auditors and inspectors, confusion or hesitation during the execution of a procedure is a red flag. Based on the trial runs, revise the SOP to eliminate the problems identified. Before you go live, test out the revised SOP again if you think it would help to further tighten up the procedure.

3.5 Final Approval

Before releasing the SOP for use, it should be finally reviewed and approved by individuals with appropriate authorities and responsibilities. The process of approval should be as per the organization's quality management plan.

3.6 Train users

Postapproval, the key to the effectiveness of an SOP lies in the training of users. The ultimate purpose of writing SOPs is to ensure the quality of processes and products, and that's possible only if the right people are trained adequately in how to use the approved SOPs.

Training is also necessary because despite having a "standard procedure," different people may still interpret it differently, which could lead to inconsistencies.

An effective SOP program includes training as one of the key steps toward the final objective, which is to ensure the quality and compliance of products and services. In fact, during inspections and audits, training often falls under the scanner.

Typically, during SOP training, a trainer explains and demonstrates why and how each step listed in an SOP is performed, and then gives learners a chance to practice. Before receiving their certification of competence, learners need to show proficiency in the understanding of the SOP.

4. SOP MAINTENANCE

Beyond writing SOPs, the next challenge lies in ensuring these documents remain valid at all times. Pharmaceutical companies operate in a dynamic environment where processes, guidelines and regulations frequently change, which in turn requires diligent maintenance of SOPs.

4.1 PERIODIC REVIEWING AND AUDITING OF SOPS

Although regulatory bodies like the FDA don't specify any time frame, ideally, SOPs should be audited three months after implementation and thereafter once every one to two years. In addition, one should audit an SOP when unexpected events like near misses or unusual quality issues arise.

A periodic audit helps to check whether employees are adhering to the procedure and whether the SOP is still valid. By implication, audits also help to identify areas for improvement—and spot opportunities to eliminate shortcomings.

Invalid SOPs should be withdrawn and archived. Whenever procedures change, related SOPs should be updated and reapproved.

Finally, the SOP auditing process should not be overly cumbersome to encourage timely and periodic reviews.

4.2 SOP labeling, tracking, storage, and archival

Typically, big organizations have hundreds of SOPs. And it is imperative that an organization maintains a master database of all SOPs, which should include these elements: SOP number, title, version number, date of release, date of last audit, author, organizational division, branch, section, and any other historical information.

An organization's quality management plan should identify the processes and individual(s) responsible for ensuring the validity of SOPs. The plan should also mention where and how the outdated SOPs will be archived. The archiving procedure should prevent continued use of outdated SOPs but make the documents readily available for historical review.

Electronic SOP storage and retrieval systems are easier to access than traditional hard-copy document systems. For the user, the electronic access should be limited to read-only format, thereby protecting SOPs against unauthorized changes. With an electronic database, automatic "Review SOP" notices can also be sent and users' training can be tracked.

Exercise (3–5 minutes)
Write three ideas to improve the current process of writing SOPs in your organization.

CHAPTER 5

Writing the Right Way

Rewriting is the essence of writing well.

—WILLIAM ZINSSER

1. Objective

IN THIS CHAPTER, we come to the heart of the matter: how to write well. What are some of the best practices in writing that can be applied to SOPs?

2. The missing piece

Where does the act of writing fit in the overall scheme of things? As discussed in Chapter 2, the goal is to write high-quality SOPs that have the following attributes:

- Well-formatted
- Concise

- Well-written
- Clear
- Precise
- Compliant

Even if you have a good technical knowledge of the procedure to be described and know about the structure (chapter 3) and the process of writing SOPs (chapter 4), one missing piece that can hold you back from achieving your goal of producing excellent SOPs is a lack of knowledge about the best practices in writing. Everyone can write, but writing well requires understanding and application of these practices. And this missing piece is often one of the main reasons behind unclear, complex, confusing, long, and ineffective SOPs.

3. Best practices for clear and concise writing

By putting yourself in users' shoes for a few minutes, you should be able to get a sense of the constraints they face while dealing with SOPs: shortage of time, too many SOPs to read, fear of non-compliance, urge to take shortcuts, impatience, and so on. How do you write in a way that helps them to understand SOPs without getting confused, frustrated, and impatient?

Here are seven best practices that can help you write clear and concise SOPs:

- Write short sentences.
- Use imperative sentences.

- Ensure parallel phrasing.
- Be specific.
- Minimize jargon.
- Use abbreviations with care.
- Write positive sentences.

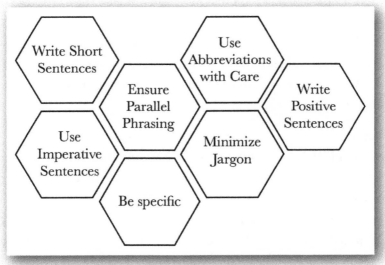

Figure 7: Best Practices in Writing

3.1 Write short sentences

In 1956, cognitive scientist George A. Miller published a paper titled "The Magic Number Seven, Plus or Minus Two: Some Limits on Our Capacity for Processing Information." Known as "Miller's Law," his research showed the number of objects an average person can hold in working memory is seven, plus or minus two. In the light of Miller's finding, you can understand why it's difficult to remember long phone numbers.

But Miller's Law also applies to the readability of sentences. Long sentences force readers to hold too many ideas in their limited working memory, leaving them tuned out, confused, and tired. Long sentences are like overloaded boats: they sink under the weight of too many ideas.

In contrast, short sentences are easier to understand since readers don't have to battle with a string of ideas in quick succession.

For example, see which of the two versions below is easier to understand and follow.

Version A
Although there are many ways to improve the quality of your writing, the easiest of all is avoiding long sentences, which are often the result of the writer trying to squeeze too many ideas into a single sentence and forgetting that our short-term memory is limited and unsuitable for absorbing multiple ideas without any gaps in between, a habit that your readers will pay for by feeling tired and confused.

Version B
Although there are many ways to improve the quality of your writing, the easiest of all is avoiding long sentences. Writers often squeeze too many ideas into a single sentence, ignoring the limitations of our short-term memory. If you write long sentences, your readers will pay for it by feeling tired and confused.

How long or how short a sentence should be? The following two guidelines provide a good framework:

- Use one sentence to describe one action or idea only.
- In general, keep the average length of the sentences to about eighteen to twenty words. Of course, this doesn't mean writing every sentence to the exact twenty-word length. Some sentences can be longer than twenty words and some shorter, but on the whole, keep the average length limited to eighteen to twenty words.

Exercise: Practice writing short sentences (5–7 minutes)
Chop the following sentences into smaller ones.

1. The study drug will be stored in a secure environment with access limited to essential research personnel, according to the storage requirements detailed in the protocol or supplied by the sponsor in a supplementary document.
2. At the request of the study monitor, drug-accountability records can be brought to the Santokba Health Care System for the monitoring visits but then will return to the SMS Cancer Specialist office since this is the site of drug storage.
3. All investigational drugs will be placed inside clear plastic bags with drug-dispensing tracking forms (attachment A) for transport and will be transported to the various sites via appropriately labeled coolers.
4. If it is outside normal working hours and doses cannot be covered from a supply of another ward, further supplies

should be ordered from the on-call pharmacist and stored in a dedicated place and locked with a unique key, which will be kept as per the key procedure.
5. Precision is generally dependent on analyte concentration and therefore should be determined at a number of concentrations, and if relevant, the relationship between precision and analyte concentration should be established.
6. Before the laboratory decides on the method and degree of validation required, the company from which the product is purchased should be contacted and all method validation information requested.
7. The procedure should initially be run using the methodology, reagents, calibrants, and controls supplied to assess the trueness of the method and allow the analyst to become familiar with the procedure.

3.2 Use imperative sentences

There are four basic types of sentences in the English language:

- Declarative: A sentence that makes a statement or tells something.
- Interrogative: A sentence that asks a question.
- Imperative: A sentence that gives a command and starts with an action verb.
- Exclamatory: A sentence that shows strong feelings or excitement.

Examples

- Declarative: You should open valve A.
- Interrogative: Should I open valve A?
- Imperative: Open valve A.
- Exclamatory: How difficult it is to open valve A!

The "procedure" section of an SOP, which directs users to act, should be written using imperative sentences. Imperative sentences give a command and tell the readers to take specific actions. Here are some examples:

Nonimperative instruction

- The dispensing valve on the container is opened.
- You should open the dispensing valve on the container.

Imperative instruction

- Open the dispensing valve on the container.

Note that imperative sentences begin with an action verb. Using the imperative form makes the instructions definite and easier to understand and follow.

Exercise: Practice imperative style (5–7 minutes)

Change the following instructions from nonimperative sentences to imperative sentences.

1. If possible, the scale should be positioned on a firm and flat surface.

2. All investigational drugs will be placed inside clear plastic bags with drug-dispensing tracking forms for transport.
3. You must fill in forms A and B to request retrieval of lost data.
4. The company from which the product is purchased should be contacted and all method validation information requested.
5. The procedure should initially be run using the methodology, reagents, calibrants, and controls supplied to assess the trueness of the method.
6. To release excess water from the tank, you should open valve A to the left of the pressure gauge.
7. Before entering the confined space, the light should be switched on.

3.3 Ensure parallel phrasing

Parallelism means using the same grammatical structure (single words, prepositional phrases, infinitive phrases, or clauses) for all ideas in a list to achieve consistency. When you write instructions in parallel form, readers don't get distracted by the form—and are able to focus more on the content. Look at the following examples:

Nonparallel: Diabetes can be affected by exercise and diet, and family history also affects it.
Parallel: Diabetes can be affected by exercise, diet, and family history.

Nonparallel

- Read the pressure.
- Once the pressure reaches the set point, close the valve.
- Record the actual pressure.
- Pump should be switched off.

Parallel

- Read the pressure.
- Close the valve once pressure reaches the set point.
- Record the actual pressure.
- Switch off the pump.

As you can see, parallel phrasing, a well-known practice of good writing, adds consistency to your content and improves readability.

Exercise: Practice parallel phrasing (5–7 minutes)

Apply parallel phrasing to the following procedure:
Procedure: Conducting Pressure Test

1. The first step is to identify the maximum test pressure to be used, as determined by the project engineer.
2. Identify the steel pipe to be tested.
3. Next, examine all connections prior to the test to ensure proper tightness.
4. Determine the pressure rating for all connected fittings and devices to ensure they are rated for the maximum test pressure.

5. Place a 150 mm diameter blind flange or other suitable cover on all openings that are not closed off by valves.
6. It's important to make sure you plug all test, drain, and vent ports that are not required for the test.
7. If the section of pipe being tested is isolated from other sections by inline valves, ensure the portion not being tested is open to the atmosphere.
8. Please apply a preliminary test pressure of 25 psi, or as directed by the project engineer.
9. Please apply the test pressure in increments of 25 psi, or as directed by the project engineer, until the maximum test pressure is reached. Hold pressure for five minutes at each 25-psi increment before adding more pressure.

3.4 Be specific

Specific and precise writing is penetrating and strong and has the power to move the readers in the correct direction. On the other hand, ambiguous writing is ineffective and weak and leaves readers confused and stranded.

While writing SOPs, be as specific as possible. Don't write, "Pull the lever" when what you mean is "Pull the lever next to the red push button." Don't write, "Refer to attachments" when what you mean is "Refer to attachment A ('List of Grade C Rooms')." Don't write, "Wait for a few minutes before switching on the particle counter" when what you mean is "Wait for a minimum of ten minutes before switching on the particle counter."

Add numbers, date, time, location, tag numbers, specific designations—anything necessary to make your writing more specific.

Exercise: Be specific (5–7 minutes)

Eliminate vague words and generalities from the following sentences, and make them more specific by adding numbers, dates, time, equipment designations, etc.

1. If you are in doubt about the correct autoclave temperature or time, refer to the appropriate SOP. (Hint: Which specific SOP?)
2. Close the door, and lock it by turning the handle. (Hint: Turn the handle clockwise or anticlockwise?)
3. Top up the salt after checking the pH value. (Hint: What's the specific limit of pH value that should trigger this action? How much salt should be added? And which specific salt?)
4. Stir the mixture, and then weigh it again. (Hint: Any duration of stirring?)
5. If the autoclave does not reach the set temperature, contact relevant personnel to arrange for servicing of the unit. (Hint: What's the set temperature? Who are the relevant personnel?)
6. Plug in the power cord, and turn the main switch to START. (Hint: Turn the switch to AUTO START or MANUAL START mode?)
7. Ensure the water container is filled up to the safety valve. (Hint: Fill it up to the top or bottom of the safety valve?)

3.5 Minimize or explain jargon

Jargon is defined as technical words or expressions used by a particular profession or group of people but which are difficult for others to understand. Dealing with jargon requires empathy with readers and a clear understanding of readers' profiles.

When writing a procedure meant only for other experts like yourself, you should use jargon or else you will frustrate your readers by defining or explaining terms that they already understand.

But if you include jargon when writing for users with varied knowledge, they may not understand certain parts of an SOP and feel frustrated or even humiliated. Such users may also include managers and top executives who may not have the requisite technical background. Here's an example:

Never transport the islet cell product directly in your hand. A secondary container or transport container must be used.

"Jargon is part ceremonial robe, part false beard," said American aphorist Mason Cooley. The following strategies might help you to save SOP users from tripping on jargon:

- Avoid using rare, difficult, or highly technical words that readers are unlikely to understand.
- If unavoidable, explain such terms the first time you use or define them in a glossary.
- If you feel defining a particular technical term not just once but every few pages would help your readers and not

force them to go back searching for its definition, please go ahead and define it more than once.

Bottom line: To keep things simple, use jargon with discretion.

3.6 Use abbreviations with care

Formed out of the first letters of words of a phrase, abbreviations (e.g., cGMP, FDA, IRS) help us avoid cluttering the content with unnecessary words. But like jargon, if used without care, abbreviations produce the opposite effect: clogging the fluidity of writing and frustrating readers.

Example
Both types of islet products (CHIP and AHIP) are cellular therapy products manufactured under GMP at the JICTF and have six hours expiration time.

CHIP, AHIP, JICTF…what are these?

Here are general rules for dealing with abbreviations:

1. For a lay audience, avoid abbreviations.
2. On the same page, include an expanded version of the abbreviation on the first use and subsequently use the abbreviation alone.
3. Consider avoiding abbreviations altogether if they appear only a few times.
4. If a document includes too many abbreviations, include an "abbreviations" glossary, which should provide expanded versions of all abbreviations.

3.7 Write positive sentences

Notice the difference between the following two sentences:

- Make sure Valve A is not open after the tank pressure reaches the set point.
- Make sure Valve A is closed after the tank pressure reaches the set point.

We can often state the same message in two ways: negatively or positively. The first sentence is negative, and the second sentence is positive.

Research confirms that readers understand more clearly and respond more quickly to instructions phrased positively. On the contrary, negative phrases are confusing, complex, and wordy.

Here are some more examples:

Negative phrasing

- Verify the green light is not off before inserting your card.
- If you don't press the yellow button, you won't be able to open the machine door.
- Don't forget to add 10 percent contingency to avoid wrong design calculation.

Positive phrasing

- Verify the green light is on before inserting your card.
- Press the yellow button to open the machine door.
- Add 10 percent contingency to ensure right design calculation.

Basically, negative sentences come in three flavors:

- Sentences that include a *no* or *not* phrase just once, but its usage is awkward, wordy, and perplexing.
- Sentences that include a *no* or *not* phrase twice—known as double negatives.
- Sentences that include a *no* or *not* phrase just once but also include some other word in the sentence that indicates a negative meaning.

Exercise: Practice positive phrasing (5–7 minutes)

Change the following sentences from negative to positive form.

1. The team spirit within the department cannot be improved if a common vision for the team is not formed.
2. Don't insert your card in the slot without seeing the green light on the display.
3. It is not recommended not to update SOPs every year.
4. If your company fails to comply with FDA regulations, you won't be able to export drugs to the United States.
5. If you don't read the SOP, you won't be able to perform the process effectively.
6. Don't forget to go through an air shower if you want to avoid contaminating the products.
7. If you don't manage the project properly, you won't be able to start the production on time.

4. Additional Writing Tips

In addition to the best practices described earlier, here are some more tips to help you further improve the quality of your writing and SOPs.

4.1 Write an informative title

In an SOP, the very first item that readers look at is the title, which should be precise and clear.

Look at the following alternative titles for the same procedure:

- Releasing a Sample During Inspection
- US Food and Drug Administration Inspection—Releasing a Sample to the US Food and Drug Administration
- Releasing a Sample to the US Food and Drug Administration

Which one is most precise and most clearly conveys the purpose of the SOP?

While deciding on a title, avoid overly long, wordy titles. A good way forward is to write a couple of alternative versions and then review each one with a simple question: Is this title clear, concise, and precise?

4.2 Write action-oriented headings and subheadings

Action-oriented headings and subheadings add to readers' clarity about the contents that follow.

For example, if instead of writing a heading like "Ingredients Weight Measurement," you write "Weighing Ingredients," it gives readers better idea of the action steps following the heading. The word "weighing" conveys action.

The technique for action-oriented headings is simple: begin a heading or subheading with verb+*ing*. Here are some examples:

- Verifying documents before transporting the sample
- Evaluating the results
- Ensuring safety of workers
- Adding reagents

4.3 Some more tips

- Avoid references to gender (use "they" or "their" and avoid "he" or "she").
- Avoid the use of "etc.," especially in running text. If the list is limited, write it out in full. If a list is too long, write the term "e.g., (for example)" and list a few items.
- Spell out the numbers one through nine. Write any larger number—10 and greater—in numerical form. Zero is usually written as a numeral (0). Above all, be consistent throughout the document.
- Clearly state what is to be done—and not what "may" or "shall" or "must" be done.
- Don't rely solely on your computer's spell check and grammar check.

- To keep things consistent, use the same term for the same item all through an SOP—not "HEPA filter" at one place and "HEPA" at another.
- When revising an existing SOP, use symbols to highlight changes so that users can quickly figure out what has changed in the latest revision.
- Use all capitals (but not too often) to emphasize importance. For example: Press the STOP button as soon as the red light comes on.

5. How much detail?

FDA or other regulatory bodies don't prescribe any specific level of detail to be included in SOPs, but a lack or excess of detail can affect the effectiveness of an SOP.

On one hand, an SOP should include all essential steps that should be performed the same way by all users. Omitting or skating over any of these essential steps may lead to confusion, eventually affecting the results. But on the other hand, procedures should not be so detailed that they are intimidating, cumbersome, and impractical for routine use. Excessive details can also cause resentment among users, thereby adversely affecting SOP compliance.

The level of detail also has an impact on audit findings during regulatory inspections. If too many details are provided in SOPs, but the same are not followed in the field, that's a huge problem. But if SOPs are too shallow, that may imply inconsistent procedures.

The guiding principle for detail is that an SOP should be detailed enough for employees to understand and carry out the activities involved. Procedure writers must include sufficient details to eliminate significant variation among users.

The level of detail required in a procedure is affected directly by the level of expertise of the individuals performing the work and the rigor of training associated with the task. In general, a higher level of detail may be required in the following situations:

- If a procedure is performed infrequently—that is, it is not a routine operation
- If different people with varying expertise are involved in the task
- If the precise execution of a procedure is critical for quality, compliance, or safety

Look at the following procedure for giving first aid to an electric-shock victim:

1. Check vital signs.
2. Establish an airway.
3. Administer external cardiac massage as needed.
4. Ventilate, if cyanotic.
5. Treat for shock.

These instructions may be clear and complete for trained paramedics but not for others. First, it includes many uncommon terms like *vital signs*, *airway*, and *cyanotic*. Second, you will notice the instructions are not complete. For example, it is not clear what kind of treatment should be given as instructed in step 5.

A procedure should clearly and precisely convey the steps to be taken in the right sequence and not leave readers guessing or using varying interpretations of instructions. The information presented should not be ambiguous or complicated. The writer should anticipate readers' likely doubts and provide explicit information.

Overall, SOP writers should often pause and ask themselves these questions:

- Are we missing any information that is required for proper implementation of this SOP?
- Are we providing any redundant information that is not really required?

Exercise (3–5 minutes)

Review the following three instructions, and decide which one includes the optimum level of detail:

Option 1

After adding ingredients A, B, and C, switch on the blender, and let the ingredients mix.

Option 2

- After adding the ingredients A, B, and C, switch on the blender.
- Operate blender for 10 minutes before switching it off.

Option 3

- Pour ingredient A into the drum of the blending machine using a 300 ml mug that has been washed before the operation.
- Wash the mug again and pour ingredient B.
- Wash the mug again and pour ingredient C.
- Start blending by pressing the green button near the lid.
- Watch the timer.
- Switch off the machine after 10 minutes.

5.1 Breaking up long SOPs

Research shows people tend to struggle with long SOPs because they cannot remember more than six to twelve steps. If an SOP goes beyond 10 steps, consider these solutions:

- Break a long SOP into several sub-SOPs.
- Write a shortened SOP that lists only the main steps without detailed explanations.
- Make the long-form document a training document or manual to supplement the shorter version of the SOP.

6. FOUR WAYS TO ORGANIZE THE STEPS OF THE PROCEDURE

Depending on the subject of an SOP, the document can include varying amounts of information with different levels of complexity. Some procedures can be simple and straightforward, while others can be long and complex.

While writing SOPs, one of the key decisions one needs to make is about the presentation or organization of the procedural steps. Considering the length and complexity of a procedure, you can choose one of or a combination of the following four ways of organizing a procedure:

- Simple steps
- Hierarchical steps
- Graphical format
- Flowchart

Two factors determine the suitability of a format for a procedure: (1) the number of decisions and (2) the number of steps.

If a procedure involves too many decision points, a flowchart is the best way to describe it. In a flowchart, you can easily show the various decision branches. But if not that many decisions are involved in the procedure, the choices boil down to using a simple format, a hierarchical format, or a graphical format. Procedures involving more than 10 steps but not many decisions are best organized in a hierarchical or graphical format. Procedures that are short and require few decisions can be best described in the form of simple steps.

6.1 Simple steps

Applicable for relatively short procedures, a format using simple steps lays out a procedure in a linear way.

6.2 Hierarchical steps

When the number of steps increases, use this format, which allows you to create a hierarchy of steps—a few main steps and then substeps within each main step.

6.3 Graphical format

When writing long procedures, one can use a graphical format, which includes images with explanatory text.

6.4 Flowcharts

Procedures that require users to make decisions should be presented as a flowchart, which is a graphical way to present the steps in a logical sequence and enable correct decisions.

The flowchart format, however, has one limitation: the level of detail should be low or else it gets too complex. Trying to include an excessive number of detailed steps can lead to a very long, messy, and hard-to-follow flowchart. Flowcharts are best used to provide an overview of a procedure while highlighting the key decisions. Some writers overcome this limitation by using a hybrid approach: using simple text-based steps to describe the procedure in combination with a flowchart. In the hybrid approach, the detailed steps can be included while keeping the flowchart sufficiently simple.

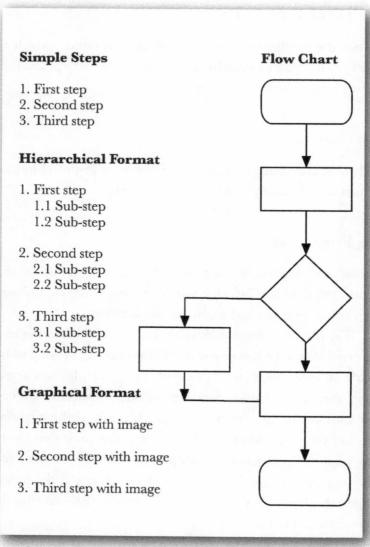

Figure 8: Four Ways of Organizing the Steps of a Procedure

7. Accessible formatting

How important is the formatting of an SOP? Does the appearance matter? If yes, why? What does the formatting of this book convey to you?

The effectiveness of an SOP is affected not just by its substance but also by its visual presentation. While busy with other tasks, readers may not be able to give their undivided attention to a procedure, and often, they may read and act at the same time. Imagine the implications if they feel intimidated by the very look of an SOP or if they miss warnings or cautionary notes or some procedural step because it is buried in a fog of dense text. Accessibility of information is crucial to the effectiveness of an SOP.

Poor formatting originates from thin margins; insufficient gaps and differentiation between key steps, subheadings, and headings; densely packed text; disconnection between content and related diagrams; inconsistent fonts; and excessive and overpowering graphical features, such as borders, underlining, and so on.

Here are some ways to make the layout of a procedure more accessible to the readers:

- **Add white spaces**
    - Use good page margins (one inch or more)
    - Leave a sufficient gap between different sections, headings, and subheadings
    - Provide sufficient gaps around notes, cautions, warning, and danger notices so these elements can stand out

- Use appropriate indentation to show hierarchy of steps
- **Line spacing:** Avoid closely packed sentences and paragraphs; use appropriate line spacing.
- **Fonts for headings:** Use appropriate fonts (type and size) to clearly differentiate between headings, subheadings, and text.
- **Font for main text:** Avoid using very small fonts for the main content.
- **Visuals:** Keep the visuals (sketches, images, flowcharts) and the related text close together. If possible, keep them side by side; if not, keep the visual right after the text. Also, set off visuals with plenty of white space.
- **Warnings:** Make warnings, cautions, and notes prominent by using bold and bigger fonts or enclosing them in boxes or both.
- **Sequence:** Arrange steps in the form of a numbered list to clearly indicate the sequence. (Use bullets if the sequence of steps is not important.)
- **Header:** Include the SOP title, number, revision number, and release date.
- **Footer:** Include the page number and the complete file name and file path to the soft copy.

Here's the bottom line: the more accessible the format, the more likely readers are to read, understand, and follow a procedure.

CHAPTER 6

Avoiding Common Errors

Good things do not come easy. The road is lined with pitfalls.

—Desi Arnaz

1. Objective

WHILE WRITING SOFTWARE code, if just a comma or bracket is placed in the wrong location, the whole program refuses to run as intended until that bug is identified and corrected. While writing a document in the English language, if just a comma or a word is placed in the wrong location, it may not stop your readers from reading, but it will probably plant some bugs in their understanding of the content.

This chapter highlights some of the more common errors in the English language. For bug-free writing, please pay special attention to the following potential pitfalls whenever you embark on writing SOPs:

- Comma splice
- Subject-verb agreement
- Parallel construction
- *Accept* versus *except*
- *Advice* versus *advise*
- *Affect* versus *effect*
- *i.e.* versus *e.g.*
- *It's* versus *its*
- *Then* versus *than*
- *Which* versus *that*
- *Lay* versus *lie*

2. COMMA SPLICE

When you join two complete sentences (independent clauses) with a comma, you commit a common grammatical error known as a "comma splice." Splicing doesn't allow readers a necessary gap between the two sentences, which amounts to something like stuffing a second spoonful of rice into your mouth before you've finished eating the first spoonful.

Examples of comma splices

- A worker and a supervisor are required to come, they have confirmed their attendance.

- Either Rashid or Anthony is attending the meeting, I am not attending.
- The SOP has been reviewed, I am going to submit it for approval.

Avoiding comma splices

You can avoid this error in three different ways:

- **Period:** You can insert a period between two independent clauses. (Example: A worker and a supervisor are required to come. They have confirmed their attendance.)
- **Semicolon:** You can insert a semicolon to signal the close relationship between two independent clauses. (Example: Either Rashid or Anthony is attending the meeting; I am not attending.)
- **Conjunction:** You can keep a comma between two sentences but add a coordinating conjunction (and, but, or, for, nor, so, or yet). (Example: The SOP has been reviewed, and I am going to submit it for approval.)

3. SUBJECT-VERB AGREEMENT

Disagreement between subject (who is taking action) and verb (the action) is one of the most common errors and affects even practiced writers of English.

Examples of subject-verb errors

- A worker and a supervisor is required to come.
- A supervisor or three workers is going to come on Sunday.
- Rashid or Anthony are attending the meeting.
- He have not finished the work.
- Either level 1 or level 2 are having this problem.

The basic point here is this: the subject (who performs the action) and the verb (the word describing the action) should agree with each other in person and number. "He writes" is fine but not "he write" or "they writes."

How to do you ensure your subjects and verbs agree? Well, just be aware of the following six situations where subject-verb agreement tends to break down.

3.1 Singular subject / plural subject

To avoid this particular error, simply remember to use a singular verb with a singular subject and a plural verb with a plural subject. Here are three examples:

Incorrect: He *have* not finished the work. (Explanation: "He" is a singular subject, but "have" is a plural verb. They don't go together.)
Correct: He *has* not finished the work.
Incorrect: A group of suppliers *are* ready to collect the tender document. (Explanation: "A group of suppliers" is a singular subject, but "are" is a plural verb. They don't go together.)

Correct: A group of suppliers *is* ready to collect the tender document.
Incorrect: Auditors *is* coming over tomorrow for the meeting. (Explanation: "Auditors" is a plural subject, but "is" is a singular verb. They don't go together.)
Correct: Auditors *are* coming over tomorrow for the meeting.

3.2 Singular subjects joined by either...or or neither...nor

When conjunctions like *or*, *either...or*, or *neither...nor* get into the stream of writing, we need to pay extra attention to avoid the following pitfalls.

Incorrect: Rashid or Anthony *are* attending the meeting. (Explanation: "Rashid or Anthony" is a singular subject—because *or* suggests only one subject will perform the action of the verb—but "are" is a plural verb. They don't go together.)
Correct: Rashid or Anthony *is* attending the meeting.
Incorrect: Either machine 1 or machine 2 *are* having this problem. (Explanation: "Either machine 1 or machine 2" is a singular subject, but "are" is a plural verb. They don't go together.)
Correct: Either machine 1 or machine 2 *is* having this problem.
Incorrect: Neither the building manager nor the owner are taking responsibility for the fire. (Explanation: "Neither the building manager nor the owner" is a singular subject, but "are" is a plural verb. They don't go together.)

Correct: Neither building manager nor the owner is taking responsibility for the fire.
Incorrect: A supervisor or three workers is likely to join the team. (Explanation: "A supervisor or three workers" is a plural subject, but "is" is a singular verb. They don't go together.)
Correct: A supervisor or three workers are likely to join the team.

3.3 SUBJECTS JOINED BY *AND*

When it comes to subject-verb disagreements, the innocent-looking *and* can play havoc.

Incorrect: A worker and a supervisor *is* required to solve the problem. (Explanation: "A worker and a supervisor" is a plural subject, but "is" is a singular verb. They don't go together.)
Correct: A worker and a supervisor *are* required to solve the problem.
Incorrect: All supervisors and the manager *is* supposed to attend the training. (Explanation: "All supervisors and the manager" is a plural subject, but "is" is a singular verb. They don't go together.)
Correct: All supervisors and the manager *are* supposed to attend the training.

3.4 TIME AND MONEY

Sometimes, dealing with time and money can derail the correctness of writing.

Incorrect: Three years *are* very long project completion time. (Explanation: "Three years" is a singular subject, and "are" is a plural verb. They don't go together.)
Correct: Three years *is* a very long project completion time.
Incorrect: Ten million dollars *are* a high price. (Explanation: "Ten million dollars" is a singular subject, and "are" is a plural verb. They don't go together.)
Correct: Ten million dollars *is* a high price.

3.5 Team and staff

Words like *team* and *staff* appear plural but are actually singular.

Incorrect: The team *are* disappointed with the announcement of incentives. (Explanation: "The team" is a singular subject, and "are" is a plural verb. They don't go together.)
Correct: The team *is* disappointed with the announcement.
Incorrect: Our staff *are* highly motivated to complete the work on time. (Explanation: "Our staff" is a singular subject, and "are" is a plural verb. They don't go together.)
Correct: Our staff *is* highly motivated to complete the work on time.

4. Parallel construction

Like in the physical world where parallel construction matters for functionality and elegance (parallel lanes on a road, the edges of a table, or rows of ceiling lights), in writing too, parallelism matters.

Whenever you list two or more items in a series, try to follow the same pattern of words for all items. Parallel construction leads to clarity, symmetry, and elegance. Here are some examples:

Incorrect: I have nothing to offer but blood, toil, tears, and sweating.
Correct: I have nothing to offer but blood, toil, tears, and sweat. (These are the historic and famous words of Winston Churchill.)
Incorrect: It is more time consuming to review and edit a report than writing the first draft.
Correct: It is more time consuming to review and edit a report than to write the first draft.
Incorrect: Some concerns about the new building are these:

- Distance from the city
- Location
- It will be expensive.

Correct: Some concerns about the new building are these:

- Distance
- Location
- Cost

The idea behind parallel construction is simple: be consistent.

5. *ACCEPT* VERSUS *EXCEPT*

Should you *accept* or *except* an SOP? The confusion between the usage of *accept* and *except* arises because the two words sound the same, but their meanings are distinct.

- Accept: to receive, admit, or agree (verb)
- Except: excluding (preposition)

Examples:

- I can't *accept* this SOP.
- In school, I liked all subjects *except* history.

6. *ADVISE* VERSUS *ADVICE*

To resolve the confusion about *advice* and *advise*, imagine you're stuck with an urgent problem at work. And to help you resolve it, a colleague of yours is offering you ideas and solutions.

Is that person *advising* or *advicing* you? Well, the act of giving suggestions is "advising," which is a verb. But the thing that you take away (suggestions, recommendations, solutions) after your meeting is *advice*, which is a noun. To sum up:

- Advise: to give helpful suggestions (verb)
- Advice: a helpful suggestion (noun)

Here's a parting shot: People who are too eager to *advise* others often don't follow their own *advice*.

7. *Affect* versus *Effect*

If Internet access is removed from your office, will you be *affected* or *effected*? As you've guessed, *affect* describes action (verb) whereas *effect* is the outcome or impact (noun) itself. The following examples show the correct usage of these two seemingly similar words.

- By giving you a tight deadline for this project report, I don't want to adversely *affect* your other commitments.
- What will be the *effect* on product sales if we start advertising aggressively?

8. *I.E.* versus *E.G.*

Both abbreviations of Latin terms, *i.e.* and *e.g.* create a lot of confusion and routinely appear in the wrong places. They both are ideally used only in parenthetical comments or footnotes.

- i.e. (*id est*) means "that is" or "in other words"
- e.g. (*example gratia*) means "for example"

The following examples should further clear the air around these two widely used terms:

- We won't be inviting any clarifications from your company (i.e., your proposal has not been shortlisted).
- There are several problems with your proposal (e.g., price is too high, scope is not clear, and delivery dates are not acceptable).

Still, is there a quick and sure way to remember the difference between *i.e.* and *e.g.*? Yes, just remember that *e.g.* starts with *e*, which stands for "example." So you will automatically use *e.g.* whenever you want to say "for example." Once *e.g.* goes in the right places, its companion *i.e.* will automatically find its own place. Both of these abbreviations should ideally only be used in parentheticals or footnotes. In running text, it's better to write out "for example" and "that is."

Final point: Both *i.e.* and *e.g.* are always followed by their common friend: a comma.

9. *IT'S* VERSUS *ITS*

The confusion between *it's* and *its* is like that typical fork in the road where turning left or right can take you to very different places. Note the difference between the usage of *it's* and *its* in the following sentences:

- Are we replacing the whole machine or only *its* parts?
- The laptop is overheating, and *it's* making an awful sound. *Its* fan doesn't seem to be working.
- I have read your report. *It's* good, but where is *its* last page?

As you can see from the above examples, "its" shows possession: Whose fan? The laptop's fan. *Its* fan. In contrast, "it's" is a contraction of *it is*. Basically, *it's* versus *its* can be captured with two simple rules:

- Remember, *it's* is simply a contraction of "it is." So use *it's* when you want to say "it is."
- Use *its* when you want to use the possessive form of "it."

10. *Then* versus *than*

Using *then* in place of *than* and vice versa is a common error that even the grammar checker on your computer will fail to spot.

Fundamentally, "then" describes a placement in time. "Than" compares two things or people. (For those who love to get technical, "then" is an adverb while "than" is a conjunction.)

Here's an example:

- We will open the packing and *then* assemble the machine.
- This new laptop is lighter *than* the one I have been using.

11. *Which* versus *that*

Both *that* and *which* are used to insert an additional clause in a sentence, but that's where their similarities end. *Which* is used when this additional clause is nonrestrictive—that is, you can get rid of that additional clause without affecting the meaning of the sentence. But *that* is used when the additional clause is restrictive, which means you can't remove the clause without distorting the meaning of the sentence.

In the following examples, you can see that the clause following *which* can be removed without affecting the meaning of the sentence. You can't do the same with the clause following *that*. Removing that clause changes the meaning of the sentence.

- Recruiters don't like to read long résumés, which is only to be expected considering they receive hundreds of such documents.
- Recruiters don't like to read résumés that are more than three pages long.

Final point: Note that a comma always precedes "which."

12. *LAY* VERSUS *LIE*

If there were an award for a pair of similar words that create maximum confusion, the pair of *lay* and *lie* would beat all other pairs hands down. *Lay* and *lie* not only have similar meanings, but they also sound and look the same in certain tenses.

LAY

Lay is a transitive verb that means to put something or someone down. Example: He lays the book on a table nearby. Note the inflections of "lay":

- Lay (simple present)
- Laying (present participle)
- Laid (simple past)
- Laid (past participle)

Lie

Lie means to rest in a horizontal position. Example: On weekends, I lie down on bed for an hour after lunch. Note the different forms of *lie*:

- Lie (simple present)
- Lying (present participle)
- Lay (simple past)
- Lain (past participle)

Essentially, *lay* is transitive, which means it requires an object, something to be acted upon; *lie* is intransitive and doesn't require an object.

13. The mighty comma

Comma, perhaps the second-most-used punctuation mark after period, often gets misused, leading to very different meanings of the same sentence. Here are some examples:

- She was carrying a light, green bag. *Or* She was carrying a light green bag.
  (The first version means her green bag was not heavy. The second version means that her bag was light green in color.)
- Let's go out and eat, Dave. *Or* Let's go out and eat Dave.
  (The first is a friendly invitation. The second implies cannibalism.)
- Although busy, Jenny decided to go to the meeting. She hoped it would end soon. *Or* Although busy Jenny

decided to go for a meeting, she hoped it would end soon.
(The first points out that Jenny was busy and went to the meeting anyway. The second is about someone who is apparently nicknamed "busy Jenny.")

Let's review the basic uses of commas.

13.1 To separate items in a series

Use commas to separate items in a series of words, phrases, or clauses.

At least three items must be present to make a series. Place a comma after each item. The Oxford (or serial) comma—the comma in a series that precedes the coordinating conjunction—is a useful way to prevent ambiguity.

- Ani, Sam, Neha, and Jane decided to attend the training course on quality management.
- The old machine squealed loudly, shook violently, and ended up breaking down.
- The instructions explained how to assemble the machine, load the software, and switch it on.

13.2 To set off introductory words, phrases, and clauses

Use a comma to set off or separate introductory words, phrases, and clauses from the main sentence.

- *Disappointed*, we left the game before it ended.
- *Expecting the worst*, we piled all the food and clothing.
- *If we plan carefully for the presentation*, we can win the client's order.

A comma is not required when a subordinate clause follows an independent clause.

- We can increase sales if we plan carefully for the grand launch of our new product.

13.3 TO SET OFF APPOSITIVES

- Rachel won first prize, *an all-expenses-paid vacation to the Bahamas*.
- Jaipur, *the capital of Rajasthan*, is one of my favorite cities.

13.4 TO SET OFF NONRESTRICTIVE/NONESSENTIAL CLAUSES

- My father, *who is still teaching*, is seventy-four years old.
- The training is in the Ascott hotel, *which is near the Clark Quay train station*.
- Singapore, *a country with a population of only 4 million in 1994*, now has a population of 5.5 million.

In the sentences above, commas enclose nonrestrictive clauses that can be removed without affecting the meaning of the main sentence.

13.5 To separate independent clauses joined by a coordinating conjunction (and, but, or, for, nor, so, yet)

- I went to bed early last night, for that's been my habit for the last thirty years.
- It's not easy to learn all the rules of grammar, and I don't think it is necessary.
- Please come on time, but I am not expecting you to rush.

13.6 To separate equally important adjectives in a series

If a series of adjectives can be joined by the word *and*, use commas between them.

- Siam avoided the friendly, talkative, pleasant man sitting next to him on the bus.
- The envelope contained two crisp, clean, brand-new dollar bills.
- This is a mysterious, scary story.

CHAPTER 7

Checking Quality with a Checklist

No wise pilot, no matter how great his talent and experience, fails to use his checklist.

—Charlie Munger

1. Objective

WITH SO MANY ideas on the table for writing high-quality SOPs—correct structure, right process, best practices in writing—how can you finally ensure the new knowledge does turn into action? How can you make sure all relevant ideas to improve the quality of SOPs do get consistently applied, regardless of who is writing the SOPs? This chapter leaves you with an effective tool to do just that: a checklist.

What's so great about having a checklist?

2. CHECKLIST CAN SAVE LIVES

One of the simplest and most low-profile documents, a checklist contains the power to save lives.

In 2006, the World Health Organization (WHO) was facing a worldwide problem of unsafe surgery. Of 230 million surgeries performed worldwide annually (2004), at least seven million people a year were left disabled, and at least one million died.

An initiative by WHO led to a 2007 conference attended by doctors, anesthesiologists, nurses, and patients from all over the world. Various options were considered to solve the unsafe-surgery problem.

- Training doctors and nurses in safe surgery practices (not implemented due to cost and logistical constraints)
- Paying incentives to doctors for performing surgery safely (not implemented due to high cost and limited impact)
- Issuing a set of official standards (not implemented due to likely limited practical application in the hospitals)

Finally, the WHO panel came up with a simple but practical and effective solution: a surgical-safety checklist.

In a pilot study, the surgical-safety checklist was prepared and implemented in eight hospitals around the globe. And postimplementation, over three months, the rate of major surgical complications fell by 36 percent and deaths fell by 47 percent.

This story of a humble checklist triumphing over the knotty problem of surgical complications comes from the book *The Checklist Manifesto* by Atul Gawande, who is an American surgeon, author, and researcher.

If a checklist can help improve surgical safety, it can certainly help improve the quality of your SOPs.

3. Checklist for reviewing SOPs

How do you review SOPs in your organization?

I have often seen people reviewing draft SOPs and giving some comments (for errors, improvements, and the like) in the first round of the review process, but then coming up with some new comments in the second round—and still new comments in the third round! And the never-ending story of writing, reviewing, and rewriting continues until the deadline for issuing an SOP looms menacingly, and there has to be a forceful closure of the review process. The key flaw in this process, which is both risky and time consuming, is that it lacks any specific checks for a systematic review. The final outcome is often not the highest quality you can attain for your SOPs.

When writing and revising SOPs, a checklist provides a consistent, systematic, and objective way to ensure nothing important is missed out in the review process. And it helps you to maintain consistently high quality of SOPs regardless of who is writing and reviewing.

Here is a draft checklist based on the ideas covered in this book. Please amend it as you like, but do try it out—and experience for yourself the power of a checklist.

QUALITY CHECKLIST: STANDARD OPERATING PROCEDURE (SOP)	
Description	**Tick (Yes/No)**
FORMAT	
1. Format: Is the SOP reader-friendly with (a) sufficient margins, (b) gaps between different sections and paragraphs, (c) distinctive fonts for text and headings, and (d) overall neat formatting?	
TECHNICAL SUBSTANCE	
2. Level of details: Are the details included in the procedure optimum—neither too much nor insufficient?	
3. Structure: Is the structure used for organizing the procedural steps appropriate (simple/hierarchical/graphical/flowchart)?	
4. Sequence: Is the sequence of steps correct?	
5. Clarity: Is the information provided complete, clear, and accurate?	

QUALITY CHECKLIST: STANDARD OPERATING PROCEDURE (SOP)	
Description	**Tick (Yes/No)**
6. Notes, cautions, warnings, and dangers: Are these included in the right places in the procedure and sufficiently highlighted with appropriate symbols and formatting?	
QUALITY OF WRITING	
7. Concise: Are the sentences short and simple?	
8. Imperative sentences: Are the instructions written in imperative style?	
9. Specific sentences: Are the sentences specific with relevant numbers, data, units, dates, tag numbers, and so on?	
10. Parallel phrasing: Is parallel phrasing used where applicable?	
11. Jargon and abbreviations: Are the key technical terms and abbreviations clearly defined?	
12. Positive phrasing: Are sentences positively phrased?	

QUALITY CHECKLIST: STANDARD OPERATING PROCEDURE (SOP)	
Description	Tick (Yes/No)
13. Title: Does the title reflect precisely what the procedure is all about?	
14. Headings and subheadings: Are they informative, succinct, and action oriented (verb+*ing*)?	
15. Readability: Overall, is the SOP easy to read and understand?	
16. Errors: Are there any grammatical or spelling errors?	
17. Numbering: Are the steps numbered correctly?	
REVIEW, TESTING AND COMPLIANCE	
18. Review: Have all relevant personnel (including actual users) reviewed the document?	
19. Field testing: Has the procedure been successfully tested in the field?	

QUALITY CHECKLIST: STANDARD OPERATING PROCEDURE (SOP)	
Description	Tick (Yes/No)
20. Regulations: Does the SOP comply with applicable standards and regulations?	

4. Seven Improvements

Based on what you've learned from this book, prepare a list of at least seven new ideas that you can use to improve the quality of SOPs in your organization.

Acknowledgements

THIS BOOK IS the result of relentless persuasion of one person who somehow believed in my ability to not just write but write a book: my wife, Manisha. For years, she has been bombarding me with "You should write a book. And I don't care what you write about, but do write." Hearing "yes, yes, yes" from me but not seeing any results, at the beginning of this year (2017), she gave me an ultimatum: "I will make sure you write a book this year."

I deeply admire her vision and determination. Thank you, dear! Without you, this child of ours would not have taken birth.

Connecting the dots backward, I also want to acknowledge the role of and thank International Society of Pharmaceutical Engineers (ISPE), Singapore. Back in 2015, ISPE, Singapore, invited me to conduct a workshop called "Writing Clear, Concise, and Effective Standard Operating Procedures (SOPs)." When I was doing the research and preparation for that workshop, I didn't know I was also setting the foundation of this book.

Finally, I thank all my friends and relatives for their encouragement and good wishes.

About Atul Mathur

ATUL MATHUR IS an engineer with more than fifteen years of experience in technical writing. He has conceptualized, written, and edited hundreds of technical documents, including case studies, guidebooks, reports, training manuals, position papers, specifications, test protocols, and procedures.

As an engineer, he has also been involved in the design, construction, and commissioning of several pharmaceutical projects.

He has been a guest speaker at several conferences and workshops organized by the International Society for Pharmaceutical Engineering (ISPE). In 2015, he was invited by ISPE to conduct a workshop on "Writing Clear, Concise and Effective SOPs."

He holds a master's degree in engineering from the Indian Institute of Technology (IIT).

Atul runs his own consulting firm, Content Alive, and lives in Singapore.

For more information, visit www.atulmathur.com

Printed in the USA
CPSIA information can be obtained
at www.ICGtesting.com
LVHW041115070424
776680LV00003B/446